ERWIN ROMMEL: PHOTOGRAPHER

VOLUME 2: ROMMEL & HIS MEN

ERWIN ROMMEL: PHOTOGRAPHER

VOLUME 2: ROMMEL & HIS MEN

by Erwin Rommel & Zita Steele

FLETCHER & CO. PUBLISHERS

NOTE: The photographs in this collection belonged to Erwin Rommel before they were seized from his widow by U.S. military forces during the final stages of World War II. Rommel, an avid photographer, intended to author another book on military strategy if he survived the war. Some of these photos he took to illustrate his strategy and maneuvers, while others were for his personal interest. To the best of my knowledge, all photos were taken by Rommel unless he is shown in a picture taken by someone else, either an unidentified German war photographer or close personal associate.

DISCLAIMER: This book does not in any way promote Nazi ideology. In addition, Rommel was never a member of the Nazi party, nor was he responsible for any war crimes or genocide.

www.fletcherpublishers.com

Erwin Rommel: Photographer
—*Volume 2: Rommel & His Men*

By Erwin Rommel & Zita Steele
Fletcher & Co. Publishers
© December 2015, Fletcher & Co. Publishers LLC.

Cataloging-in-Publication data for this book is available from the Library of Congress.
Library of Congress Catalog Number 2015959643

Photography: Erwin Rommel
Author, Editor & Illustrator: Zita Steele
Interior design: Noël Fletcher
Photos of Zita Steele by Noël Fletcher
First Edition
Published in the United States of America

Cataloging information
ISBN-10 1-941184-09-X
ISBN-13 978-1-941184-09-7

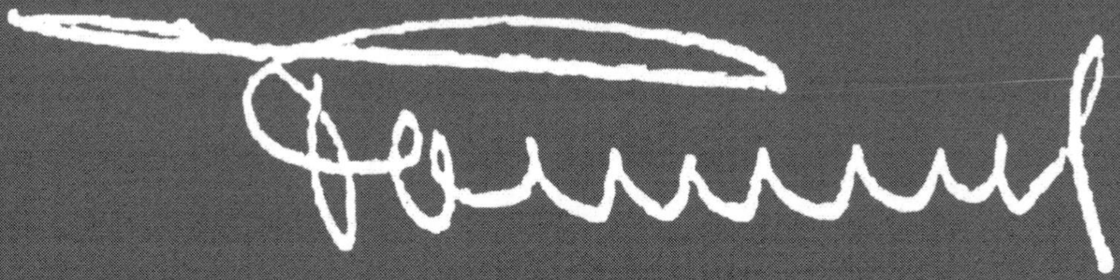

Contents

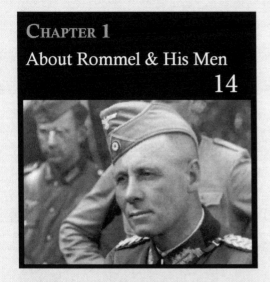
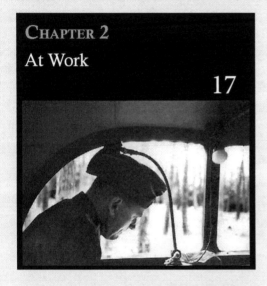
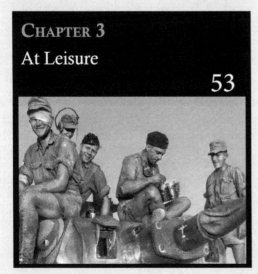

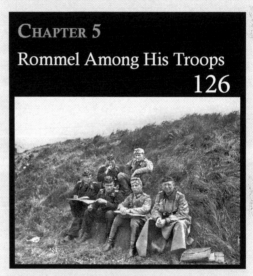

THIS SERIES

Photography was one of Field Marshal Erwin Rommel's favorite pastimes. Between 1940 and 1942, he took thousands of photos. These images cover a wide variety of subjects reflecting both Rommel's military and personal interests. Rommel was meticulous in labeling his photos. He intended many of these pictures to be illustrations in a manual on military tactics that he planned to write after World War II. His untimely death prevented him from doing so.

Rommel also collected pictures of himself taken by German war photographers and others. Rommel evidently saved the images to document or remember important events in his life. These photos are interspersed throughout each book in the series.

My goal in the *"Erwin Rommel: Photographer"* series is to provide readers an opportunity to gain insights about Rommel through both the photos he took and those he collected. The photos are presented to provoke thought and allow people to make their own interpretations. The series offers a unique perspective on Rommel, while providing a glimpse of what he experienced and saw in the war.

The first book, *"Volume I: A Survey,"* contains a broad overview of Rommel's photographic work and a sampling of overarching themes in his collection. The other books explore individual themes in greater detail.

VOLUME 2: ROMMEL & HIS MEN

Rommel's photo collection reveals that he took great interest in his fellow soldiers. Rommel literally took hundreds of photos of his comrades. Rommel also saved numerous pictures of himself among his aides and troops; he likely preserved these images as mementos.

These images provide a more personal glimpse of Rommel, as a common soldier rather than a general. The photos have little military emphasis. They are candid snapshots that depict ordinary soldiers interacting or engaging in various activities.

The pictures of Rommel himself are unpolished. There is a noteworthy lack of glamorous propaganda photos. Rommel often appears dusty and sunburnt as he attends casual events, mingles with other soldiers, or takes turns using his camera to take group photos with his staff.

The subject matter reveals what specific moments that Rommel wished to capture: images of men at work, at leisure, on the move, and Rommel spending time with his troops and comrades. (The themes are categorized in the chapters of this book.)

Many of the photos from Europe were taken in Belgium and France in 1940. The photos from North Africa were taken in 1941 and the early half of 1942.

MY APPROACH TO THESE PHOTOS

As a graphic artist and writer, I am well aware of the power and intrinsic ability of images to convey information using visual elements that are stronger than printed words. Images, however, can be distorted through various means, such as cropping and design manipulations. In this manner, photographs, especially in this digital age, can be taken out of context, altered from the original image, or slanted to show a different viewpoint contrary to the intent of the photographer.

One of the most important goals of my restoration was to preserve the photographer's original work. The contents within the images have not been distorted in any way.

All photos taken by Rommel are shown in full frame, as they actually appeared when he captured the shot. For some photos, I have extracted close-up sections and placed them next to Rommel's full-frame images so readers can see interesting details I observed in certain areas within the pictures.

The only photos which have been cropped are those taken by others of Rommel himself. This was done to make it easier for viewers to see Rommel clearly, particularly in pictures of him surrounded by other people. In some cases, I have extracted close-ups of Rommel, cropped them to provide greater visibility of him, and placed them next to the original photos of himself that he collected.

Rommel used a Leica camera for much of his photography. While the exact model is unknown, the Leica III D camera was popular during World War II. Rommel's camera and accessories were stolen after his death from his home in Germany by American soldiers, who even took his uniforms.

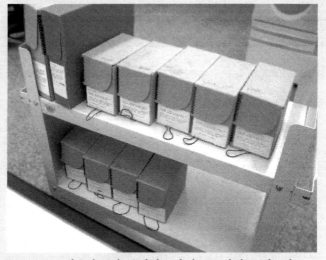

To create this book, I delved through hundreds of photos, slides, negatives, and photographic contact prints in these boxes at the National Archives where Rommel's personal photos are stored after having been seized by American forces during World War II.

ROMMEL'S IMAGES & MY PHOTO RESTORATION

The Counterintelligence Corps of the U.S. military seized Rommel's photo collection in 1945 from his widow in Germany. They confiscated his Leica equipment and about 3000 photos that he had taken. The images, which fill about a dozen boxes in the National Archives, are in a variety of conditions. Many are jammed together in stacks so tightly that some images are stuck to others. Some are slides encased in plastic sleeves. Others are large contact prints that contain series of images from film negatives. Most however, are still photos printed on various types of paper and in various sizes.

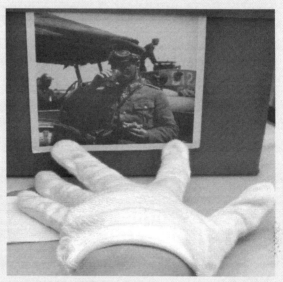

While examining the photos at the National Archives, I place a picture on the table while wearing cloth gloves.

The majority of the photographs in Rommel's collection are badly damaged. In some cases, the photos are stained, yellowed or darkened by age. Many are creased and slightly torn. Other photos contain significant amounts of scratching, dust, and blotches due to the conditions in which they were taken and developed.

I am responsible for the restoration of all photos in this book. My graphic design and multimedia skills are central to my work as an author and artist. All the photos in this book have undergone a painstaking digital-restoration process. I used a three-phased approach to the restoration. The process is described below.

■ *Repaired and Lightened*—Because the photos were all very dark, they had to be lightened in order to be visible. Scratches, dust particles, tears, stains, and other forms of damage were repaired.

■ *Balanced*—Due to poor lighting conditions that occurred when many of the photos were taken, not all key elements within the photos were equally apparent. Therefore, I balanced the toning and lighting as needed to revive the pictures.

■ *Revived*—I made a special effort to bring out all the details in each image. I wanted to bring the pictures to life by making each aspect of the subjects as clear as possible.

Three examples of my restoration work in the various phases are shown below.

REPAIRED & LIGHTENED

Original photo (left). The same photo after my restoration (right).

BALANCED

Rommel's original photo (left). The same photo after my restoration (right).

REVIVED

Original photo (left). The same photo after my restoration (right).

ROMMEL'S HANDWRITING

All of Rommel's handwriting shown in this book is authentic.

The handwritten words shown on the Chapter title pages were chosen for artistic effect; their meaning in German is relevant to the Chapters before which they appear. All of these were taken from Rommel's own photo captions and writings.

Some of Rommel's writing is shown throughout the Chapters. Captions may appear next to the photos on which they were originally written or photos containing relevant subjects.

Rommel's signature shown at the beginning of the book is from an item in my private collection.

A sample of one of Rommel's handwritten captions on his photos.

1

About Rommel

&

His Men

Gen. Rommel

Rommel had a close relationship with his men. As someone who began his career as an infantryman in World War I, he felt a close affinity with fellow soldiers, particularly frontline troops. He viewed them as comrades rather than subordinates. These sentiments are revealed in his personal letters and writings. For example, Rommel wrote that his men were "very dear" to him.

He was attentive to the men's general mood as well as details about the lives of the rank and file soldiers. Despite the demands of his position as a commander, Rommel often took time to visit ordinary soldiers and talk with them on a personal level. He also shared treats with them that people sent to him as gifts. He refused extra privileges and lived the lifestyle of a regular soldier on the front.

Rommel also showed concern for his staff officers. In his personal letters home, he often discussed how his staff officers were doing—such as discussing their health and shared activities. In many letters, he described events in a group sense using the terms "we" and "us" to denote both himself and his men.

He noted individual bravery in his memoirs, often paying tribute to soldiers who demonstrated particular courage or leadership. Although many of his writings focused on military events, Rommel also recorded anecdotes about the adventures he experienced with his staff officers—memorable events he clearly wanted to preserve.

Rommel made no distinction between the different ethnicities of soldiers under his command. Whereas many German commanders at that time only praised the qualities of German troops, Rommel showed no partiality; his writings contain no nationalistic sentiments. Rather, he referred to the German troops with understated pride and wrote more detailed accolades about many different types of soldiers. For example, he praised Italian, Australian, New Zealand, and British combatants.

He maintained relationships with his comrades from both WWI and WWII. After the defeat and surrender of the Afrika Korps, Rommel wrote letters to his men in prisoner of war camps. He also made efforts to find out where particular acquaintances were interned in order to correspond with them. In letters to his wife, Lucie, he referred to his men as "the brave boys" and said he was depressed at having been relieved of his command prior to their surrender. Instead, he expressed the wish that he could have stayed with the troops on the battlefield in North Africa.

In turn, Rommel was much loved by the troops under his command. The men not only esteemed Rommel as a military leader, but they viewed him as their friend. During WWII, Rommel's men demonstrated great loyalty to him. Among themselves, some soldiers in North Africa referred to him as "Erwin" rather than by his surname or military title. For many years after the end of WWII, former Afrika Korps soldiers held an annual memorial service at Rommel's grave and left wreaths in tribute to him.

In contrast to his other photography, Rommel rarely took pictures of his men from the air. When it came to documenting life among his men, his photography was up close and personal. The expressions on the faces of many soldiers show they are looking straight at Rommel as he photographed them, often smiling in an interaction with him behind the lens.

These photos that Rommel took and collected give a glimpse of the experiences that he and his men shared as common soldiers: as they worked, rested, traveled, and prepared for battles together. The majority of these photos have no military significance and appear to have been taken as mementos. The images give the reader a unique perspective about the scenes of soldier life that Rommel wished to capture as a photographer, and what people and moments he wanted to preserve in time.

2

At Work

Kvass

Major Otto Heidkämper, Rommel's chief of staff in 1940, looks over a map while seated in a car during winter in Europe (date and location unknown).

Rommel is shown next to his two assistants as they discuss military strategies.

After photographing Heidkämper seated next to him in the back of a car, Rommel passed the camera over to another occupant in the car. Since he was planning to write a book after World War II on his operations, it is likely Rommel intended to publish these photos to show him and his soldiers at work in the field.

Two German soldiers crouch on the ground next to an artillery gun covered with leaves.

The soldiers and their gear are centered in the frame. The men appear to be looking into the distance through a clearing in the trees. The thick foliage indicates Rommel took this photo during military operations in the summer at an unknown place in Europe, likely France.

Rommel noted the brave actions of two of his men during a dangerous skirmish in France.

"But with enemy shells falling all around and a hail of splinters whistling around our ears, it was not easy to get the infantry to leave cover...Lieutenant Corporal Heidenreich and my driver Lieutenant Corporal König particularly distinguished themselves by their cool courage in this situation. They swept the infantry forward..."

Rommel (third from left) joins a group of soldiers moving an object hidden from view, while other soldiers (right) carry a metal frame. A closeup of Rommel (right) is extracted from the photo.

Rommel looks to the right as his men continue moving the frame.

Although the object being moved in the top photo is obscured, thick metal chains can be seen (lower, center). Rommel took a very hands-on approach to leadership, as shown in these photos.

A Panzer is moved from a ship in North Africa to the ground below (above and below).

Both German soldiers and sailors pull ropes to maneuver the Panzer over the side of the vessel. The facial expression of one sailor (lower right) highlights the great physical effort needed to pull the rope under the heavy weight of the steel tank.

The scene around the tank shows German troops involved in a variety of activities.

The men work to release the Panzer, which is marked with an "I".

Soldiers (left) help to hoist another tank from the ship to the ground.

An African dockworker walks in the center of a road as two Italian officers stand with their arms folded (right).

Amid the bustle of activity at the port, most people on the street show no interest in a jeep as it hangs in a net suspended from a vessel. Rather than focusing on the jeep, Rommel's photo conveys the busy atmosphere of a port in wartime.

A young German sailor (right) gives direct eye contact to the camera as
Rommel takes the photo near a Panzer bearing the number "144".

Men stand atop a building (left) overlooking a port where an ambulance is
being lowered from a ship.

Libyan dockworkers help unload German vehicles onto barges in North Africa.

The viewpoint of this image, with ropes at the bottom of the frame, shows Rommel stood on the edge of a vessel when he took the photo. The vehicle on the barge is in the center, balanced by workers on both sides. Symmetry and a strong focal point are characteristic features in Rommel's photo compositions.

German sailors look over the ship (top right) in Rommel's direction as he captures a moment of relaxation from dockworkers.

A barefoot man (center) lights a cigarette as his colleagues joke. The ropes at their feet indicate all the vehicles have been unloaded from the ships.

German and Italian troops arrange fuel canisters on a beach.

Diverging curves are a strong focal point. Rows of men (left) move in a semicircle across the beach.

Soldiers collect floating supplies, which seem to be fuel canisters, off the North African coast.

One man (center) stands on a rock formation and looks towards the water as a vessel cruises nearby.

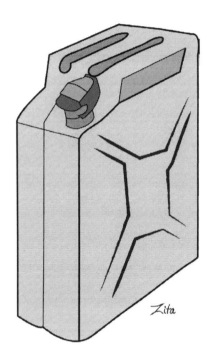

German fuel canisters, such as those shown in Rommel's photos, were a revolutionary invention. First produced in Germany in the late 1930s, these steel canisters (called *Wehrmacht-Einheitskanister*) were robust, portable, and could hold 5.3 gallons of gasoline each.

Compact and equipped with three handles, they were easy for soldiers to carry and transport. They were specially built to be leakproof and designed to prevent fume combustion. Since these cans were all-in-one units, no tools were necessary to open or use them.

These German canisters were viewed as a phenomenon when first seen by Allied forces during WWII. Both British and American troops used flawed tins to transport gasoline; they were heavy, flimsy, difficult to carry, and impossible to open without the use of various complex tools.

British troops in North Africa, seeing the ingenuity of the German-engineered containers, began to collect German canisters from the battlefield for their own use. The canisters acquired the name "jerry cans." The German design was copied by Allied forces and mass-produced. Today, jerry cans are manufactured and used all over the world to hold fuel and other substances.

A German in a pith helmet and an Italian with goggles over his cap (left) look at Rommel as he takes a closeup photo of soldiers filling fuel canisters from metal drums.

Fuel was of critical importance in the North African theater. Both motorized weapons and transport vehicles depended on gasoline to function. Rommel wrote, "Troops have to shoot and move, and for that they need ammunition and petrol." Lack of supplies, especially fuel, ultimately caused Axis forces to lose the War in North Africa.

British intelligence intercepted Rommel's requests for supplies using the Enigma machine. The British strategically targeted and sunk German and Italian supply ships to North Africa, cutting off Rommel's source of fuel and ammunition. This forced the retreat and ultimate collapse of the Afrika Korps. Rommel's writings reflect this truth.

> *"Retreat to A_____ [Agedabia]. You can't imagine what it's like. Hoping to get the bulk of my force through and make a stand somewhere. Little ammunition and fuel, no air support. Quite the reverse with the enemy."*

> *"Some very hard days behind me. We had to break off the offensive for supply reasons and because of the superiority of the enemy air force—although the victory was otherwise ours."*

Rommel's troops stand around the fueling station in North Africa.

Rommel stood closer to the action (top) and stepped back for a wider view of his men in both photos. Metal drums could easily become searing hot in the sun.

Throughout his time in North Africa, Rommel and his men suffered in the harsh desert.

"It was 107 degrees here yesterday, and that's quite some heat. Tanks standing in the sun go up to as much as 160 degrees, which is too hot to touch."

"A quite atrocious heat, even during the night. One lies in bed, tossing and turning and dripping with sweat."

"I usually spend a lot of time traveling, yesterday I was away 8 hours. You can hardly imagine what a thirst one gets up after such a journey."

"The heat's frightful, night time as well as day time."

Surrounded by ropes, a German officer looks up from sunken ground.

The angle of the photo indicates Rommel was looking down when he snapped this picture, which appears to be related to the photo (below) of a rope tied around a crumbling wall.

A German officer watches a Libyan lead a horse, saddled with a barrel.

A mosque (upper left) stands among buildings in the distance.

Various views of a drum-laden camel in a camp with tribesmen and soldiers.

Rommel must have been very interested in the camel judging from the different angles of the photos. Undoubtedly conveying military supplies on camels was an exotic experience for the German commander. The white minaret of a mosque appears in the top and bottom photos.

A German soldier in a pith helmet stands as sentry, facing the outside of a building.

Rommel composed this photo of the lone soldier balanced in the center using a frame-within-a-frame concept. The soldier stands under a door frame, and the camera lens is positioned outside the ledge of an interior doorway.

A solitary soldier sits on a rock pile near a tripod as he scans the horizon from a desert outpost.

Another German soldier keeps watch from a dugout.

Rommel took both of these photos from a vehicle as he monitored frontline activities. The exterior of his vehicle (bottom) and another passenger (right) appear in both images.

Rommel's soldiers sit in position near heavy artillery on top of a ridge.

The serious expressions and outward gaze of the soldiers indicate they were on the lookout for enemy troops. The metal shield on the top of their helmets, which had crown openings for ventilation, bore the black, silver, and red national colors of Germany.

The morale of the Afrika Korps was very high in the beginning of North Africa campaign and during the duration of their successes. Rommel captured this confidence felt by himself and his staff in his writings.

"We looked forward to the battle full of optimism, trusting in our troops, with their superb tactical training and their experience in improvisation."

Two soldiers dig sandy trenches as an Italian officer watches for the enemy.

The inner wall of the trench forms a dominant curving line in the image. The men are located along a coastline. The sea is visible at the top of the photo. A pair of military binoculars rest on the sand next to the Italian officer.

Rommel had many close calls during various battles in Europe and North Africa. He described one harrowing incident.

"Between 10 and 12 we were bombed no less than six times by British aircraft.

On one occasion I only just had time to throw myself into a slit trench before the

bombs fell. A spade lying on the soil beside the trench was pierced clean through

by an 8-inch splinter and the red hot metal fragment fell beside me in the trench."

Dozens of soldiers shovel sandy soil to create fortifications in North Africa.

Roadwork was necessary to allow the movement of Rommel's troops and supplies. In his writings, Rommel noted the difficulties he and his men faced due to lack of roads in North Africa. Although the Italian government had created colonies in North Africa, it did not build many highways. There was a lack of paved routes connecting one place to another. German supply trucks often broke down driving on rough dirt roads. Rommel requested Italian cooperation to build roads, but no help was forthcoming. As a result, ordinary German and Italian soldiers were tasked with road construction.

In this photograph, the shoveling men and path form parallel diagonal lines that diminish in size (from left to right) into infinity. Rommel's photographs often contain strong diagonal elements in the composition. The men's clothing show the image was taken in cooler weather compared to other photos, which depict bare-chested soldiers in shorts working beneath the intense desert sun.

Soldiers work to level a path for vehicles to travel through the desert.

Deep ruts of sand in the lower forefront of the image provide an indication of driving difficulties. The vehicle (middle left) has just passed over this road and is parked beyond a low ridge.

A large group of soldiers conduct various work around a roadway.

The high vantage point of this image indicates Rommel took it from atop his command vehicle on the road. The center road dominates the photo as it disappears into infinity. Small clusters of soldiers walk into the desert (upper right).

A German soldier in a greatcoat stands alongside a roadway next to a vehicle and troops.

This image contains familiar themes in Rommel's work: a prominent lone figure (the profile of the soldier in the center) and diagonal lines (the road and horizon). At the bottom of the frame (next to tire tracks) is the shadow of Rommel's vehicle.

German and Italian soldiers look up at the camera as Rommel travels along the curving road.

The men sitting and standing by the side of the road appear to be taking a break. In the middle right, one soldier cuts the hair of another, who is seated with white cloth covering his shoulders. Clusters of soldiers are in the distance near a convoy truck (upper left).

A bareheaded soldier (center) pauses from digging with a shovel as his photo is taken.

These Italian soldiers are fortifying a rock wall. One soldier carries a stone (left) in front of another who holds a pickaxe. The rock wall cuts a diagonal line through the image.

A pair of Italian soldiers lift shovels with heavy sand as they create a desert trench.

An empty tin can sits next to a tanned German soldier working in a trench.

A soldier carries a shovel load of dark earth from an opening he digs in a hillside.

Rommel's shadow appears in the lower right of both images as he captures photos of his men at work.

A shirtless German soldier raises a sledgehammer to crack rocks near a
battered shovel above a trench.

The caps of two men (left) are barely visible as they stand inside the trench. In the center of the image.

Rommel was sensitive to the hardships his men faced in every aspect of their lives in North Africa.

"The physical demands on the troops during this period approached the limits of endurance. This placed a particular duty on the officers to provide a continual example and model for their men."

A mechanic in stained shorts (center) looks down next to a vehicle in a makeshift repair shop, while another soldier crouches next to a motorcycle (right).

The Germans built many workshops in North Africa. These were used to repair machinery, including vehicles, equipment, and weapons. Rommel made frequent use of these workshops. He used them to construct his so-called "Cardboard Division" —decoy tanks which fooled British reconnaissance.

On one occasion, the use of German workshops created friction between Rommel and Italian commanders. During an inspection of a desert area, Rommel found weapons in the sand that had been abandoned by Italian troops during a battle a year earlier. Since the guns were still serviceable, Rommel had them repaired at German workshops and put into use on the frontlines.

The Italian High Command was displeased. It notified Rommel that the weapons were Italian property and for Italian use only. Rommel wrote indignantly, "They had been perfectly content up until this point to stand by and watch this material go to wrack and ruin, but the moment the first guns had been made serviceable on our initiative, they began to take notice. However, I was not to be put off." The restored weapons, which included antitank guns, were used to strengthen the Afrika Korps defenses.

Dust clouds surround the barrel of a long-range gun as soldiers practice shooting.

Three men crouch to position a gun as a crowd gathers.

Several men hold binoculars while staring out into the direction where the gun barrel is pointed. Two soldiers near the base of the gun wear long knives. One man (center, left) appears to hold ammunition in his hands. The uniforms of many men are dusty and stained.

A man wipes his brow (left) while his two comrades talk beside a large piece of artillery.

The crumbling building in the background, with a man staring from a doorway, seems to be a workshop area with machinery scattered around.

A bearded Italian soldier with goggles on his head (center) looks down from a rugged hilltop as his photo is taken.

The men are working on trench, which is indicated by the pickaxe resting on top of the ground (upper left).

Sand embankments provide cover for men and artillery (right).

A multitude of footprints in the sand form a patterned texture that dominates the forefront of this image.

A German soldier trudges in the sand away from a gunner, who reclines at his post.

Two German soldiers smile in conversation as one man reads a document.

This asymmetrical positioning of the two men in the far left corner of the frame conveys an interesting balance. The men are located diagonally in the opposite direction of a hilltop that looms over them on the top right. The central focal point in the image is the curving terrain that dips in the middle ground and background. The composition of this photo contrasts the enormity of nature against man.

Rommel often wrote about his adventures in North Africa, some of which were humorous. While seeking a delayed Panzer division, he experienced the following.

"To form my own judgment of the situation I drove forward to the leading scouts to see what was happening. I found them lying in a cactus grove close beside an Arab village. Heavy artillery fire was falling in the village and confusion was complete, with every living creature, bird and beast, scattering in all directions. Bayerlein collected up eggs which some hens had dropped. Then we, too, had to get into cover, and Bayerlein crawled amongst the cacti carrying his precious booty. We came to no harm—and neither, fortunately, did the eggs."

A canteen is cast aside on a ledge as a German soldier emerges from a tunnel dug into a hillside.

A central lone figure is a feature common in Rommel's compositions. Rather than take a closeup, he included all the surrounding natural elements.

Five soldiers clear out rocks from a tunnel they are digging.

The small wheelbarrow indicates they had to make multiple trips lugging heavy rocks away. The angle of both photos suggests Rommel stood on the ground above and cast his lens down when he took the photos.

Rommel (left) stands in conversation with three German officers.

Rommel was excited to arrive in Africa. Like most German soldiers, he had exotic impressions of the country. (See Page 152.) He enjoyed being outdoors and exploring new places. The people, landscapes and animals of North Africa were the subject of many of his photos. Although Rommel spent much time in the desert, he occasionally took opportunities to sightsee.

The hot, dry climate was also beneficial for Rommel, who suffered from arthritis. His physician previously advised him to travel to Egypt for some relief. Rommel frequently wrote about appreciating the sun and the warm weather. In a letter to home, he expressed the idea that it would be nice to travel through Africa in peacetime.

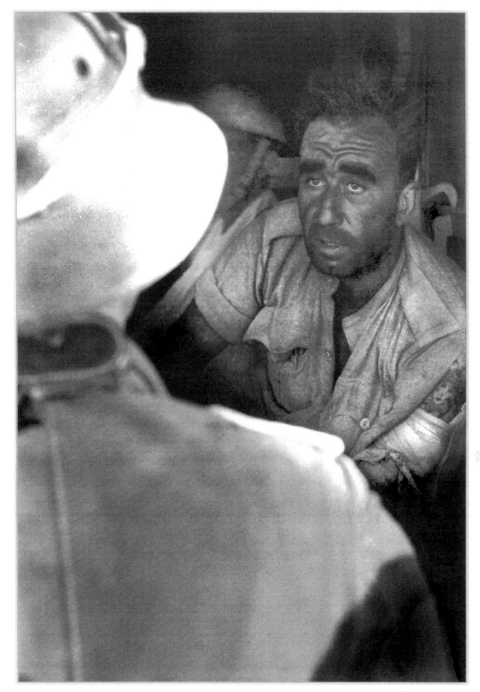

A weary Italian soldier listens to a troop member wearing a pith helmet.

Rommel's shadow comes into view (lower right) as he stood behind a soldier to capture the exhausted expression on the face of the Italian, whose disheveled hair is standing up. The man's face and clothes are dirty. A Harlequin tattoo peers out above the bandage on his left forearm.

Two young German soldiers trudge through the hot sand carrying supplies on their shoulders.

One soldier (left) smirks as Rommel takes the photo. Their rumpled uniforms and discolored boots are worn out. Their military helmets differ from other troops in North Africa, who wore cloth caps and pith helmets. On the horizon are a multitude of vehicles and a pair of soldiers (upper left).

3

At Leisure

Akropolis

Rommel's troops pause to grin for the camera while relaxing on top of a tank.

The soldier (seated on the left) has a bandage around his head and right arm. Its shirtsleeve is torn above the arm wound. His legs, arms, high boots, and clothing are covered in dirt. Behind him, a soldier wearing headphones grins as another leans from behind to smile broadly for the photo with his tin cup at his side. In the center, an Italian soldier wearing goggles around his neck sits cross-legged while eating from scratched metal container. The Italian and the man standing next to him (far right) both wear rings. Two pith helmets are fastened to the top of the tank (left). An image of an airplane is painted on the gun barrel. This photo shows both the camaraderie among Rommel's men and the hardships of war.

Rommel wrote about how he and his men wanted the war to end.

• *March 12, 1942* •

"We're all hoping that we'll be able to bring the war to an end this year. It will soon have lasted three full years."

Two soldiers converse in a busy shipyard filled with various German troops.

A set of tracks forms a strong curving element in this crowded port scene. Soldiers occupy both sides of the frame as well as the center. Numerous soldiers stand around doorways next to a building (left). On the opposite side, troops are in the middle of the street. Some appear to assist with the unloading of a vessel while men outside the building look on. This image is balanced on both sides of the frame by pairs of soldiers engaged in conversation. The two on the right face each other and stand as if oblivious to the action around them. The men on the left both have bowed heads as they travel slowly down the street. The man on the bicycle holds his balance with one foot as he chats with his companion, who ambles down the road.

A passenger (left) in Rommel's Storch airplane smiles for the camera while the pilot (right) looks out from the cockpit.

Rommel's reflection taking the photo can be seen in the window (left) next to the passenger.

Rommel (second from left) sits among his men and appears to be speaking to them.

Rommel's Iron Cross is barely visible beneath the lapels of his jacket. A pair of goggles rest on the brim of his cap. His men have serious expressions as they listen.

Italian and German soldiers in greatcoats mingle outside a building.

Some soldiers carry documents. A gun holster is visible on the back of one man (right).

A soldier sits with his legs draped over a low wall along the North African coast.

The shadow is visible (left) of the building where Rommel stood to capture this image from a high vantage point.

While at leisure, Rommel took this photo of a harbor scene.

The solitary man (left) is a familiar element in Rommel's photography. The roadway, palm trees, and ships form horizontal elements in the composition.

A truck (left) leaves clouds of dust in its wake as it races towards a crumbling concrete divider, where a lone German soldier stands with his hands behind his back.

Rommel took this photo from a hilltop looking downwards at the coast. The curving road upon which the truck travels and a lower road (center right) form parallel curving elements in the composition. A single man against a large backdrop is another signature characteristic of Rommel's photography. He could have centered the image on the soldier, but instead pointed his lens in such a way that the lone figure contrasts the vast distance of the sea on the right side of the frame. Another feature of Rommel's work is the use of dominant diagonal lines. In this image, the striped concrete barrier juts into the frame.

The man in the photo is dressed in warm clothing. Rommel wrote about the extreme weather.

"I visit the troops every day, most of them are by the sea. Occasionally we bathe. The water is still very warm, and the heat pretty bad during the day, but it's so cool at night now that I need two blankets."

Soldiers relax while listening to one of their comrades sing and play the guitar.

Men stand atop a vehicle in a crowd that gathers to enjoy the entertainment.

Many soldiers grin as they listen to the music. One man in a helmet with glasses turns from the group to watch Rommel take their photo.

Rommel was an avid hunter, who decorated his walls with trophies and enjoyed eating wild game. He continued to hunt near the battlefront in North Africa. He and his officers often hunted gazelles for food. They would use their cars to chase the gazelles and shoot with pistols and rifles from moving vehicles. He records two such events.

"I went out shooting last evening with Major von Mellenthin and Lieutenant Schmidt. It was most exciting. Finally, I got a running gazelle from the car. We had the liver for dinner and it was delicious."

"...our Christmas dinner trotted up to us in the shape of a herd of gazelles. Ambruster [an interpreter] and I each succeeded in bringing down one of these speedy animals from the moving cars...

"At about 17.00 hours General Bayerlein and I joined the H.Q. Company's Christmas party, where I received a present of a miniature petrol drum, containing, instead of petrol, a pound or two of captured coffee...At 20.00 hours I invited several people from my immediate staff to share a meal off the gazelles we had brought in that morning."

A pair of soldiers (center) stand next to a metal drum inside an old Arabic fortress.

The equipment and military presence there may indicate Rommel used the fortress as a base for his operations. He took the photo of the interior from a high angle.

Two desert tribesmen on horseback pass through a crowded street.

The soldiers in the crowd watch the travelers with interest. Two men (right) hold cameras next to a man wearing a fez.

Five members of Rommel's troops pose in front of his Mammoth command vehicle, which bears the Afrika Korps palm tree emblem and has a gun barrel protruding form the front window.

Hans-Joachim Schräpler, Rommel's adjutant, is second from left and wearing a peaked service cap. He accompanied Rommel during his campaigns in France and North Africa (See Pages 137 and 139.) Several months after arriving in North Africa, Schräpler was killed in an accident; he was run over by the Mammoth, one of Rommel's command vehicles. In this photo, Schräpler stands beside the very vehicle that would later take his life. He left behind a widow and two young sons, one of whom later published a book containing Schräpler's letters from North Africa.

Rommel wrote about how his subordinates frequently fell out of action due to hazards in the desert.

> "Now Gause is unfit for tropical service and has to go away for six months. Things are also not looking too good with Westphal, he's got liver trouble. Lieut. Col von Mellenthin is leaving today with amoebic dysentery. One of the divisional commanders was wounded yesterday, so that every divisional commander and the Corps Commander have been changed within 10 days."

> "My commanding officers are ill—those who aren't dead or wounded."

A single soldier stands in the middle of a desert plain surrounded by others who lay on the ground behind rock barriers that provide cover against enemy fire.

Battles in the desert were very difficult due to the harsh natural environment and the stress of fighting. There were many hardships, particularly since battles could last for many days at a time. Rommel described one such situation.

"The battle is still going on and will need all our efforts if we are to win it.

Prospects are good, but the troops are dead tired after 12 days of it.

I'm in good form, very lively and ready for anything."

Four Italians in a gun crew smile broadly for the camera with their rifles and gear scattered around them.

One soldier (upper left) sleeps face down in the dirt among weeds near his comrades. Rommel liked to visit his troops at the frontlines. Following a decisive victory in North Africa, Rommel traveled around the battlefield to personally congratulate them. In letters to his wife, he described this experience.

• *June 18, 1941* •

"The three-day battle has ended in complete victory. I'm going to go around the troops today to thank them and issue orders."

• *June 23, 1941* •

"I've been three days on the road going around the battlefield. The joy of the 'Afrika' troops over this latest victory is tremendous..."

A colonial soldier looks at Rommel as another climbs from an odd shelter on the beach.

Italy colonized Libya in the early 1900s. As in many colonized nations, there were conflicts between the settlers and native peoples. Many Italians immigrated to Libya over the course of decades. Mussolini's fascist government attempted to unify the diverse population under Italian rule. Many native North Africans joined the Italian military as colonial troops. The men in this photo were members of the Libyan Carabinieri, a division of the Italian military police.

Rommel (center, left) stands on a rock listening as one of his soldiers opens a document for discussion.

Rommel appears to be at an outdoor while the unknown photographer stands near a desert bush and snaps the picture.

Rommel (center) turns left to speak to German and Italian officers during a tour of ancient ruins in Cyrene in North Africa.

The photographer of this image captured another photographer (bottom) who was documenting this outing among the military leadership.

Several soldiers explore the Acropolis in Athens.

Rommel toured ancient ruins in North Africa. It is likely he made a brief stop in Athens where he took this photo of the Acropolis. He clearly enjoyed traveling and exploring noteworthy sites, which appear in his photos. In his memoirs, Rommel recounted a memorable experience exploring a ruin with his staff in January 1943.

"We took this opportunity to visit the old Roman city of Leptis Magna, the ruins of which were still standing. An Italian professor acted as our guide and explained the features of the place in excellent German. But our thoughts were more with Montgomery than the ancient ruins. Moreover, the strain and lack of sleep of the past few days were beginning to tell, and my A.D.C., Lieut. Von Hardtdegen, particularly distinguished himself by falling asleep between two pieces of feminine statuary. Bayerlein photographed him there."

A net covers the exterior of a field kitchen where two soldiers relax.

A soldier (center) leans against a supporting pole on the field kitchen and smiles. Another soldier sits bent over a folding seat that is being used as a table, holding a cup. Containers and other articles are located inside the truck. In the distance, a figure (left) of a man walks away from the eating area.

Rommel's tent stands on a road next to his command vehicle, metal drums, and assorted rubble.

Rommel was constantly on the move in North Africa and did not always have proper living accommodations. At times, he lived in his car or slept in the back of trucks. Sometimes he set up his headquarters in tents or small buildings. The Italians gifted him a van, which he used as his sleeping quarters whenever possible. He once wrote that he considered it a luxury to have a roof over his head.

In this photo, a wooden door for privacy is propped up against a pole. The opening reveals the humble dwelling were Rommel rested from his hectic duties as the commander. In the privacy of his tent, he wrote daily letters to his wife Lucie about his day and other matters. Sometimes even sleeping in a tent was a luxury.

"I've been living in my car for days and have had no time to leave the battlefield in the evenings."

"One loses all idea of time here...I've been up at the front area for a few days, living in the car or a hole in the ground."

Rommel photographed his accommodations inside his tent in his headquarters near Tobruk.

Wooden planks form a floor above the sand. A hammer is placed next to a supporting pole. A scarf is used as a tablecloth. It appears to be Rommel's scarf, gifted by his daughter Gertrud. This tent appears to offer little protection against the harsh desert elements.

An alternative view of Rommel's tent.

This angle of his tent shows how neat and orderly he was even while living in makeshift accommodations out in the field. A telephone receiver is on the table within easy reach next to the large folding chair where he sat.

When the German campaign in North Africa faltered, Rommel was under enormous pressure to succeed when the tide had turned against him and his men. He expressed his crisis to his wife.

"The battle is going very heavily against us...At night I lie open-eyed, wracking my brains for a way out of this plight for my poor troops. We are facing very difficult days, perhaps the most difficult that a man can undergo. The dead are lucky, it's all over for them. I think of you constantly with heartfelt love and gratitude. Perhaps all will yet be well and we shall see each other again."

Rommel (center left) poses for a group photo with nearly 40 of his men while attending an Easter party in 1941.

The lively atmosphere of the Easter celebration is apparent as the men enjoy posing for the camera. The group consists of Italian and German soldiers. They crowd around Rommel and even peer over his shoulder as they try to squeeze into view.

A close-up of Rommel in the crowd.

This photo has been cropped and enlarged to provide a better view of Rommel enjoying himself with his men.

Rommel (center) turns to the left as a wider view is shown of him with his men.

More soldiers can be seen smiling for the camera around their commander. The car on the right is covered in sand and dust. Rommel appears to be wearing a cloth Italian cap in this image.

A close-up of Rommel, extracted from the image above, shows him turning to the side.

He is holding some unknown object in is hand, possibly his German Army service cap, which he usually wore.

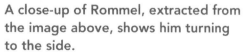

Six soldiers sit together in the sand during a break as smoke rises in the distance.

The men in the back of the row gaze at a tank and two vehicles in the distance. One soldier (left), with his back to the camera, sits on a box. His gear includes two knives and a canteen. A few soldiers smoke cigarettes. The man closest (right) to the camera sits with his hands clasped and has a stoic expression as Rommel snaps the photo.

Major Otto Heidkämper holds a document, looks out from his monocle, and smiles.

Heidkämper was Rommel's chief of staff. (See Page 18.) While in France, Heidkämper criticized Rommel's innovative command methods; he complained to administrative officers and sent Rommel a document outlining his grievances.

Rommel was angered by Heidkämper's actions and wrote to his wife that he needed to "put the boy in his place." However, the two men were on fairly good terms, and the issue was soon resolved.

"I'm glad there's peace in the camp again," wrote Rommel afterwards.

After serving in France, Heidkämper spent the remainder of the WWII fighting in Russia and on the Eastern Front. He received the Knight's Cross. He was captured by American troops in 1945 and released soon afterwards.

In this photo, Heidkämper stands in front of a brick building. The tree branches are bare.

Rommel takes a drink while eating a sandwich during a break from operations in France.

This is a rather unusual photo for Rommel to have kept in his private collection. He is depicted in the middle of eating a meal with his half-eaten sandwich in prominent view. In the background is a soldier on top of a Panzer (right). Rommel's uniform has a series of holes above the breast pocket, left there by military decorations pinned to the cloth. A French soldier is visible behind Rommel on the left.

Rommel and his officers share a meal at an unknown open-air field kitchen in Europe.

Rommel and the other men nearby stand as they eat. The plates have been placed on a makeshift table. Rommel focuses on his food as an officer on his right holds documents and appears to be talking. In the background, men stand around. One soldier is seated on the grass next to vehicles (right).

Rommel firmly believed his place was among his men. He described this sentiment in his memoirs.

> *"The commander must have contact with his men. He must be capable of thinking and feeling with them. The soldier must have confidence in him. There is one cardinal principle which must always be remembered: one must never make a show of false emotions to one's men. The ordinary soldier has a surprisingly good nose for what is true and what false."*

A close-up of Rommel from above.

Rommel takes an impromptu photo of five traveling companions in an airplane.

A fatigued German trooper (center) lifts his helmet and leans against his motorcycle.

The soldier is located between an industrial site next to metal scraps and a crane. The front half of a Panzer (right) is parked next to the crane. Three military trucks with camouflage tarps are in the street. As in his other images, Rommel composes this photo with a solitary figure set against an imposing backdrop, in which machinery and buildings tower over the man.

Rommel's men surround a tank on the beach in France.

Two soldiers walk on the beach.

A diagonal line of footprints crosses the sand ahead of the man in the center.

German troops in France explore their beach surroundings along the Normandy coast.

The man in the center holds documents tucked under his arm and a case in his hand as he turns from the seascape. Next to him, a soldier looks through a pair of binoculars beyond the coastline.

In the background is a Panzer marked on the side of its turret. A crewman wearing suspenders has climbed atop the tank to retrieve something among the various pieces of gear and equipment.

Dozens of German soldiers relax on a grassy hill overlooking the sea along the French coastline.

Everyone is seated on the ground except an officer (right) who has papers on his lap and has removed his hat. The officer appears to be resting on a folding chair. Two young soldiers in the center look at Rommel as he snaps the photo. Next to them, a hatless soldier looks through binoculars towards the horizon over the sea.

Rommel believed strongly that military leaders should set a good example for their subordinates. He frequently wrote about this philosophy.

"From my officers, I demanded the utmost self-denial and a continual personal example, and, as a result, the army had a magnificent esprit de corps.

"The experience of this magnificent and entirely spontaneous loyalty between officers and men kept hope alive even through the darkest hours of the African war. Even in Tunis, the troops retained full confidence in their command— probably a unique phenomenon after a retreat of 1,200 miles. But a bitter fate denied them any escape to Europe."

Rommel is seated in the center in a grassy field among six of his aides.

Rommel sits straight up with his boots crossed at the ankle and looks at something that is being pointed out by a soldier (center left). Two soldiers recline (far left and center) with documents set on the grass next to them. The man (center) standing with binoculars is one of four wearing black Panzer uniforms. Several tanks are visible in the background. Two tank crews (upper left) rest on the ground in front of their vehicles. One Panzer is marked "R12".

A close-up has been taken from the photo of Rommel gazing from a high vantage point.

A haystack provides cover for Rommel and two of his soldiers.

Rommel sits on the haystack with his hand casually placed inside his pocket. Two other soldiers stand atop a vehicle where straw has been placed over it as camouflage.

Bombs and shellings caused many difficulties for Rommel throughout the war. Rommel lost many German colleagues, who often had no time to find cover.

"We dived straight for cover in a trench to our right, but not before a shell had killed the dispatch rider, Ehrmann, and wounded the signals officers, and N.C.O., and a second dispatch rider."

4

On the Move

nach der ägypt. Grenze.

Rommel believed in being leading from the front, particularly when his troops were on the move. He explained his viewpoint on this subject in his memoirs.

"It is also greatly in the commander's own interest to have a personal picture of the front and a clear idea of the problems his subordinates are having to face...If he fights his battles as a game of chess, he will become rigidly fixed in academic theory and admiration of his own ideas. Success comes most readily to the commander whose ideas...can develop freely from the conditions around him."

A touring car and motorcycle trooper watch a motorcycle column speed through the countryside.

A herd of cows dot the field (right) and ignore the commotion around them. One cow (center) curiously approaches the road. Three of the speeding motorcycles have sidecars. The convertible touring car is filled with other soldiers. Two buckets hang from the rear bumper.

In his military operations, Rommel did his best to create strategies that would minimize casualties. In a letter to his wife about fighting in France, Rommel expressed gratitude for the low number of casualties among his men.

> *"A day or two without action has done a lot of good. The division has lost up to date 27 officers killed and 33 wounded, and 1,500 men dead and wounded. That's about 12 percent casualties...The worst is now well over."*

He later wrote the following.

> *"The best form of 'welfare' for troops is first-class training, for this saves unnecessary casualties."*

Crewmen on three Panzers travel across a field as other ground troops follow.

The soldier in the center hoists a rifle over his shoulder. A dark cloud of smoke drifts from the direction the men left.

Soldiers holding rifles rush through a field with three tanks.

The slight blur on the back of one tank (right) indicates Rommel took this photo as the Panzers were moving.

Rommel (second from left) and his aides, standing on a higher elevation, look tense and have serious expressions during military operations in a field.

Soldiers clamber around an idle vehicle (right). Inside another car (center) a driver waits. It is likely this was Rommel's car for himself and his aides.

A group of soldiers stand around a motorcycle and tanks.

A circle and diamond shape are visible on two tanks (center). Trash is scattered on the ground near the motorcycle.

Over a dozen soldiers climb up a hillside to hurriedly cross railroad tracks in a rural area in France.

Rommel must have been moving with his men and climbing up the hill to monitor developments when he stood between the train tracks to take this photo. Some soldiers in the middle of the tracks look like they are placing flat objects (right) while others seem to be pulling the wooden railroad ties out from the tracks. The soldier in the center carries a grenade launcher.

Snipers frequently targeted the Germans as they advanced through France. Rommel noted in his writings that one of his men was wounded.

"Snipers were still very active, mainly from the left, and a number of men had been hit, including Lieutenant von Enkefort, though his was no more than a graze."

A group of soldiers crouch on the ground around a grenade launcher.

Most of the men have moved past the railroad tracks and ascend a steep hill (right).

A pair of Panzers pulls a disabled tank from a gully.

One Panzer marked with the number 131 (left) and another each have cables tied to them as they attempt to pull another tank (marked 112). Its wheels appear to be stuck on the bank of a gully. A soldier holds onto the turret and rides atop the immobile tank. A few other men stand around watching (upper left and right). The viewpoint of this image indicates Rommel was positioned on top of a hill when he took the photo. This suggests the tank was descending from a steep hill and attempting to cross the gully when it became stuck.

A German officer stands next to a tank (above) sinking in the mud. Two cables are attached to the front of the tank. Then the officer climbs on the tank (below) to assist as it is pulled further from the mud.

A German officer stands on a riverbank as an armored vehicle starts to cross a makeshift bridge.

A group of soldiers stand in icy water watching the vehicle move past them.

One man in the water adjusts a beam under the prefabricated wooden sections of bridge as the vehicle drives by into a snow-packed forest.

Two soldiers guide a wooden section through the water as other soldiers congregate in a forest as a truck drives over the bridge.

More troops ride (above and below) in a vehicle towing an antitank gun cross the bridge.

Other soldiers and artillery are transported across the river while soldiers in the water monitor the bridge (above and below).

German soldiers move from fields and a street into a town.

The camera lens is directed at the unfolding scene from Rommel's vantage point as he traveled down the road close behind his men. The soldiers scatter across the area (right) and can be seen walking through tall vegetation and tree-lined fields. A brick church (left) is the only building in view.

As Rommel traveled down a road in France, he encountered a young German soldier and wrote about him in his memoirs.

"As we drove back along the main Dieppe-Paris road, we passed a German tankman bringing in a French tractor with a tank trailing behind it. The young soldier's face was radiant, full of joy at his success."

Clouds of dirt follow tanks (center) driving up a road to join a Panzer formation (left).

A cameraman films a tank crew in action while the gunner looks through binoculars.

An assistant holds the legs of the tripod steady as Rommel captured this ironic image.

A motorcycle trooper (left) accompanies a convoy through a city.

Rommel must have turned around inside his vehicle to document the scene behind him as the combat units speed down a street. An officer stands with his head sticking out of a car as it travels through the city. No pedestrians are visible.

A squad of six motorcycle troopers (above) follow a car and two armored vehicles down a street.

Rifles are slung over the backs of the troopers. The sides of their motorcycles are packed with gear.

Four motorcycle troopers (left) pause along a street with a large presence of military vehicles in the background.

Rommel's motorcycle troopers drive along a brick street in Paris in 1940.

Rommel took this photo while attending the *Siegesparade* (Victory Parade) in Paris. (See Pages 134–135.) He evidently crouched down low to take this photo from the opposite side of the street. The parade was an occasion of great pomp and ceremony for the German armed forces. All high-ranking members of the Third Reich government were present and passed through the streets among the marching troops. Amid the nationalistic spectacle, it is interesting to note that Rommel only saved three photos of the event: this picture of his troops passing by, and two images of him standing at attention and saluting his men.

Above man waves a flag (center left) in the middle of the street. Rommel structured this image by holding his camera at an angle to capture a series of diagonal lines. The street, trolley tracks, and vehicles form dominant horizontal lines across the foreground and middle. He also shows the motorcyclists gradually diminish in view from the right to the left, which is also characteristic of his compositions.

Tanks roll down a street while other vehicles park on sidewalks to make way.

Small groups (above) of German soldiers gather in various locations along the street. Many are watching the tanks.

A destroyed tank.

A soldier turns away from a broken down tank to face Rommel as this image is taken. Debris is scattered in front of the battered and scorched tank.

A small dog (left) watches as three goats join a platoon march and trot behind two soldiers (center) on bicycles.

Rommel undoubtedly saw the humor in this scene when he snapped this photo. The blur of the bicycle wheels and the goats' legs show they were moving down a street at a fast pace. Only the officer on the horse (center, right) notices Rommel with his camera. The soldiers who march in the platoon carry a variety of gear and weapons. Many of them converse with each other. Both the men on the march and the goats ignore each other. The trio of goats are focused on following the soldiers on the bicycles.

An officer (center) walks in a street in a military parade as soldiers ride atop vehicles.

Italian soldiers in a car, with their national flag painted on the hood, await departure.

A man walks to another vehicle with soldiers. The convoy of cars is parked in front of a derelict building.

Benghazi April 41

A soldier approaches three men in a vehicle at the Atiq Mosque in Maydan al-Huriya, Benghazi in April 1941.

The mosque looms in the background. Two soldiers sit on the upper frame of the vehicle. Only the head of the driver appears above the steering wheel. In the distance, two Libyan men (left) walk down an isolated dirt street.

Another view of the Atiq Mosque in Benghazi.

Rommel stepped around the car in the previous photo. A vehicle (right) drives up to the mosque and a soldier opens the car door as this photo was taken.

A soldier in a pith helmet (left) stands sentry at the open gates of a fortress.

The curving road in the center of the frame dominates this image. Again Rommel uses a familiar composition technique of contrasting a lone figure with an immense backdrop. On the wall of the fort (right) above a doorway is a shield emblem with an eagle. The the vehicle (right) with the open door is probably the truck Rommel was riding in when he stepped into the road to take this photo.

German troops parade after arriving in Tripoli in a photo series.

Rommel waves to men (center left). Rommel used a military trick to deceive audiences at this event.

Rommel (center) watches his troops drive past him and other military leaders.

In order to deceive Allied spies, Rommel ordered the German tanks at the parade to drive past him multiple times. The tank crews took an elaborate route through alleys in Tripoli and then reemerged onto the main thoroughfare, driving in a continuous circle.

To bystanders on the main streets, this gave the illusion of a massive German military force. Newspaper articles around the world reported that large array of German troops had arrived on African shores.

In reality, Rommel's Afrika Korps was small division at that time. The British were also deceived by these reports, which caused them to overreact to Rommel's first attacks and lose territory in battle.

Rommel is surrounded (above and below) by many Italian commanders.

Rommel has taken both the top and center profile photographs. He aimed the lens higher in the top image to incorporate more of the desert backdrop.

Then he shifted the angle of the camera to show more of the officers in their long coats. One officer (left) holds onto the strap of a pair of binoculars, while the other wears a leather glove on his left hand.

The man with binoculars exchanges places with Rommel in the bottom image. Rommel poses with his right hand on his hip as he and the other officer face the camera.

Rommel and a group of officers take turns in front of the camera on an airfield.

A German and Italian officer (top) stand near an airplane as a crew of men and airplane workers (left) watch. Rommel takes both the top and center photos.

Another officer (left) joins the group. The tail of the airplane becomes visible. Its pilot (center, left) stands near the wing tip.

Rommel has passed his camera to another. Rommel stands with his hands behind his back (center) among more officers, who have come into view. Most of the soldiers have departed. An tribesman (left) watches from the back.

An Italian motorcyclist speeds past Rommel through the desert in the first of a series of photos.

The motorcyclist has a rifle slung over his left shoulder. The back of his bike is packed with gear. He doesn't appear to be wearing any goggles to shield his eyes. The texture of the sandy soil is rippled with lumps and small vegetation. The composition of this photo is very balanced. The sky and the sandy soil occupy almost equal halves within the frame. The lone figure of the man and machine are centered.

Rommel praised the qualities of Italian soldiers in his memoirs.

"There were splendid Italian officers who made tremendous efforts to sustain their men's morale. Navarrini (XXI Italian Army Corps) for example, for whom I had the highest regard, did everything he could."

"The duties of comradeship, for me particularly as their Commander in Chief, compel me to state unequivocally that the defeats which the Italian formations suffered...were not the fault of the Italian soldier. The Italian was willing, unselfish and a good comrade...Many Italian generals and officers won our admiration both as men and soldiers."

Eight Italian Bersaglieri motorcycle riders are shown in a diagonal line as they disappear from view into infinity (above and below).

A sedan is parked across the street from an intersection showing the way to Tripoli.

The edge of a building (right) indicates Rommel stood outside a structure when he photographed activity near the sign across the street. That parked car could have been provided his transportation. A variety of military vehicles are parked in the distance. Several soldiers surround the vehicles. Two trucks (center) drive towards the intersection.

Although the Luftwaffe provide air support, Rommel's supply trucks and other vehicles were often targeted by British aircraft.

"Driving sandstorms blew on and off all day, making the lives of my men a misery—although they did, at the same time, prevent the British air forces from making any heavy attacks..."

Dust clouds follow two vehicles carrying soldiers across the desert to an area with activity (right).

A soldier smiles (left) next to vehicles filled with Rommel's troops.

A driver (center) looks at Rommel through a sand-covered windshield. Vehicles fill the frame. Several soldiers (left) in one dirty vehicle watch as their picture is taken.

Seven soldiers (left) adjust an antitank gun as several vehicles (center) drive away from a site where black smoke swirls across the sky.

Three men stand on the vehicle to position the artillery. More than a dozen large trucks in the distance (center and right) head away from the battle scene towards the position where the artillery is being set up. The profile of a tank (center) sits on the horizon next to the churning smoke.

Rommel's letters to his wife often were frank when describing the circumstances that he and his troops faced.

"We're launching a decisive attack today. It will be hard, but I have full confidence that my army will win it. After all, they all know what battle means…I intend to demand of myself the same as I expect from each of my officers and men.

My thoughts, especially in these hours of decision, are often with you."

Soldiers riding in the back of trucks are buffeted by fierce wind in a desert
storm, leaving enormous plumes of dust in their wake.

Rommel took this photo from his airplane; the tip appears in the upper right. One
man in the truck waves at him.

Rommel peers out of his airplane during a reconnaissance mission to take a
photo of his men digging trenches.

The rows of men form a diagonal pattern. One shirtless man (center) waves his hand
at Rommel's plane as it passes.

A group of soldiers stand inside the back of a dirt-covered truck as it travels on a road away from camp.

The men in the truck are perfectly centered within the frame. The road forms a curving feature from the front of the image to the side. In the distance, other soldiers walk through a desert campsite. The angle of this image indicates Rommel took the photo from his command vehicle as it traveled down the road ahead of the group of men in this photo.

Traveling in the desert always resulted in clouds of dust. Rommel described the appearance of dust as his troops moved across the terrain.

"Columns were rolling eastwards along the track raising great pillars of dust."

Soldiers stand on a makeshift bridge looking at an ambulance's undercarriage.

A man's hand on a spade (left) comes into view while his shadow appears on the opposite side of a trench. The construction of the bridge and its supporting beams is intricate. The shirts of some men are soaked with perspiration.

Comrades carry a wounded soldier from battle to safety and medical treatment.

A medic (right) tends to the wounds of the injured soldier.

The man has injuries on his face, arm, leg, and foot, including burns and cuts. These photos of the injured soldier are significant because Rommel had no other photos of the wounded in his personal photograph collection. He could have captured and saved these images of this man and saved them due to some act of bravery that the soldier did in the field.

In his writings, Rommel reflected on the cost of human life in the forsaken desert.

"Rivers of blood were poured out over miserable strips of land which, in normal times, not even the poorest Arab would have bothered his head about."

5

Rommel
Among
His Troops

A weary-looking Rommel, with goggles across his forehead, (center) sits on a hillside with his assistants and Panzer crewmen.

Rommel collected these photos of himself with his troops for personal mementos. Rather than stylized or distorted propaganda photos, the pictures in Rommel's private photo collection were never intended for publication. The images capture moments in time as he traversed Europe and the Middle East with his soldiers.

Rommel points to the left as his subordinates are engrossed by a map.

Around Rommel's neck are his camera strap and binoculars. The men stand in a field covered with wildflowers. In the background stands a Panzer. The face of one tank crewman appears above Rommel's hat.

Rommel peers through binoculars as his assistant opens a map in a grassy field.

Rommel was energetic during his advance through France. It was his first time commanding tanks in battle. His high spirits was evident in a letter to his wife.

"Everything wonderful so far. Am way ahead of my neighbors. I'm completely hoarse from orders and shouting. Had a bare 3 hours' sleep and an occasional meal."

Clutching his camera in his left hand, Rommel smiles as he waits with a group of soldiers.

One soldier (left) holds a flare pistol pointed at the ground. A convertible touring car is parked behind the men.

Rommel leans against a chair (left) during a meeting with other officers, who study maps outside a house in Europe.

The speed of Rommel's advance through France was unprecedented and led to his division earning the nickname "Ghost Division." He wrote of his successes to his wife.

"Two glorious days in pursuit, first south, then southwest. A roaring success. 45 miles yesterday."

Rommel (center) appears to be enjoying himself as he paddles a boat across the Meuse River in France.

Lt. Most (left) served alongside Rommel in many dangerous situations. Rommel admired him and was deeply moved when Most was killed in action in May 1940 in France. Rommel noted an act of bravery from one of his men as they crossed the river.

"Rubber boats paddled backwards and forwards and brought back the wounded from the west bank. One man who fell out of his boat on the way grabbed hold of the ferry rope and was dragged underwater through the Meuse. He was rescued by Private Heidenreich, who dived in and brought him to the bank."

Rommel (left) turns to listen to a Panzer officer (right) whose hand is blurred in the foreground.

The other Panzer crewmen appear to be in deep thought as they sit in the grass during a break.

Rommel (left) has stepped out towards the street and stands in salute.

Other German officers stand along the street (right and center) during the ceremony.

Wearing a steel helmet, Rommel (center, right) stands at attention in France.

This closeup image of Rommel has been extracted from the original photo.

Rommel preserved these photos of himself at the Paris Victory Parade, watching and saluting his men. During this event, he found time to snap a photo of his troops as they passed him by on the opposite side of the street. (See Page 104.)

Rommel (right) and one of his aides stand with their arms behind their backs beside a brick building.

Rommel's face looks haggard and he appears to be unwell. This could be a result of fatigue. During the campaign in France in 1940, Rommel often commanded on the frontlines for several days at a time with no sleep.

Rommel (left) studies a map held by Schräpler, who wears a Condor Legion tank badge on his pocket as he briefs a group outside the town square of St. Valery, France.

Rommel (center) and other men turn as an officer with a cigarette holder points out something beyond the view of the French garage in the background.

Rommel stands next to French Commanding General Marcel Ihler of the 9th Army Corps in St. Valery-en-Caux (1940). British General Robert Fortune stands nearby (right).

This is one of many photos that Rommel preserved of the capture and surrender of Allied forces at St. Valery-en-Caux, France. This was a decisive German victory in the Battle of France and, at that time, one of Rommel's greatest military achievements.

The German young man without a cap (center, right) received a special mention Rommel's account of the victory that day.

"A German Luftwaffe lieutenant, who had just been liberated from captivity, was made responsible for the guard [of the captured generals]. He was visibly delighted by the change of role."

Rommel (right) congratulates his staff following victory at St. Valery. Major Heidkämper (center) nods as he shakes Rommel's hand. Schräpler salutes.

Rommel (center) converses with an officer in front of a large building.

With his hands in the pockets of his great coat, Rommel looks relaxed like the other man who rests his hands on his hips. Two other soldiers (left and center) look bored as if awaiting transportation.

Rommel stands on a rocky hill (center) talking to his aides, some of whom carry documents.

The photographer sneaked this photo of Rommel while hidden alongside a tent in foliage. This impromptu image apparently captured Rommel with his mouth open talking in mid-sentence.

A military broadcaster conducts an audio interview of Rommel in North Africa.

Rommel's expression is one of reflection while the interviewer pauses. The audio technician sits in the car monitoring the recording. Rommel's goggles and boots are caked with dirt.

Rommel (right) leans against a rail of his command vehicle in the late afternoon.

Rommel's clothes show heavy perspiration as do those of the soldiers around him. A vehicle with Italian soldiers (left) travels opposite.

In his writings, Rommel described the appearance of a main road through the desert.

"The Via Balbia stretched away like a black thread through the desolate landscape, through which neither tree nor bush could be seen as far as the eye could reach."

This is a closeup of Rommel from the above photo.

Rommel pins a military decoration on an Italian officer as his other troops look on.

The German soldiers in pith helmets smile during the ceremony. Rommel's RV sleeping van is parked nearby. Rommel mention decorating a soldier named Von Wechmar with the Knight's Cross.

A wider view of the ceremony reveals two more Italian soldiers standing beside to the others in the photo above.

Italian general Italo Gariboldi prepares to award Rommel (left) with a military decoration in this photo series.

A calendar on the wall (left) shows that the date is the 22nd of April.

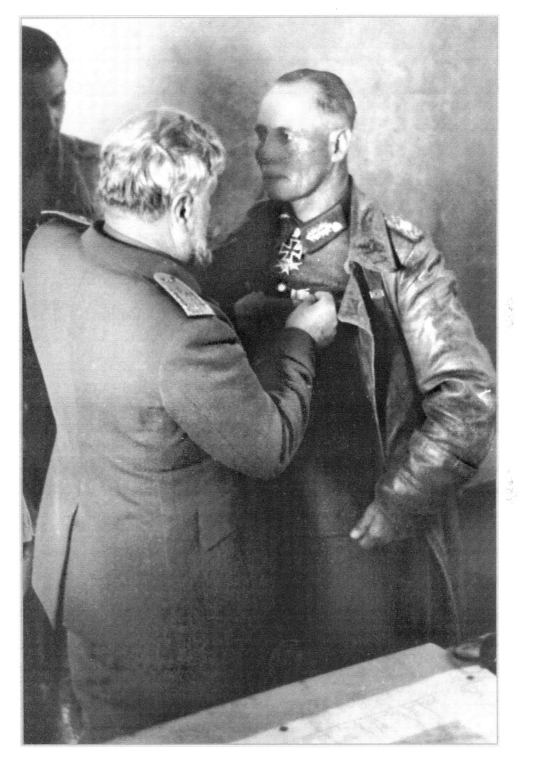

Rommel appears serious as Minister Attilio Teruzzi pins the award to his uniform.

The complexion on the upper portion of Rommel's face and head (uncovered by a hat) is a pale compared to the tan caused by the desert sun. His leather coat is worn and battered.

On April 22, 1941, during a day of intense fighting in front of Tobruk, Italian commanders Italo Gariboldi, Mario Roatta, and Attilio Teruzzi, Minister of Italian Africa, visited Rommel to present him with an award.

Rommel, who described the battle situation as "highly critical," was presented with the Italian Medal for Bravery. He also was told that he would receive the Italian equivalent of the Pour le Mérite.

Unlike German medals, Italian medals were usually large and very ornamental. While German award ceremonies tended to be brief and minimalistic, Italian leadership bestowed military honors with great formality. As can be seen in these images, Italian decorations were often given with an embrace—another feature uncharacteristic of German award ceremonies.

The Medal for Bravery was one of several Italian decorations Rommel received in North Africa. When he heard in April 1942 that commander Ettore Bastico was coming to present him with yet another Italian award, Rommel wrote the following.

"I can't say I'm terribly thrilled about it. More troops would suit me better."

Rommel later recounted the ceremony in a letter to his wife.

"He presented me, in the name of the King, with the new Colonial Order. A large silver star, even bigger than the previous one, plus a red sash with a small order.... This really is enough."

General Gariboldi speaks to Rommel (above) and congratulates him (right).

Binoculars are around Rommel's neck in the photo above.

After the military decoration is awarded, Rommel smiles but has a weary expression on his face. His tan line is more apparent across the front of his face.

Teruzzi embraces Rommel after decorating him.

Rommel (center) smiles broadly as he meets with the Italian crew of a ship in this photo series.

The rugged North African coastline appears in the background (above and below).

Rommel and the crew stand together in different poses.

Rommel travels in a small boat with three members of the German Navy.

Rommel was encamped briefly near the coast of the Mediterranean Sea in Libya. He enjoyed his time there and wrote about in letters home.

"Spent our first day by the sea. It's a very lovely place and it's as good as being in a hotel in my comfortable van. Bathe in the sea in the mornings, it's already beautifully warm. Aldinger and Günther living in a tent close by. We make coffee in the mornings in our own kitchen."

Rommel (left) strolls through ruins with a group of German soldiers led by Gariboldi.

A closeup is taken of Rommel (center) among the tour group.

Rommel carries a book as he meets with German and Italian officers.

The book in Rommel's hand is *"Unvergessenes Kamerun: Zehn Jahre Wanderungen and Jagden 1928 -1938,"* by Ernst A. Zwilling, first published in Germany in 1939. It described the author's 10 years of experience of safaris and exploration in Cameroon, Central Africa.

Rommel read nonfiction books to improve his knowledge of foreign countries. It is possible he bought this memoir to prepare himself for travels in Africa. Like most German soldiers, Rommel seems to have had exotic impressions of the African continent.

However, Cameroon, bordered by Nigeria, has an extremely different landscape and climate from North African countries such as Libya, Egypt, and Tunisia. If Rommel hoped to gain insight about his surroundings by reading this book, it seems likely he did not find it very helpful.

In this photo, most of the soldiers are wearing uniforms with shorts except Rommel. An Italian soldier (left) wears his socks rolled down to his ankles.

Rommel (center) strides past soldiers as General Gariboldi guides him and another German officer.

Rommel's tall leather boots and dark jacket are more formal in his military uniform than the shorts worn in the desert.

When Rommel first arrived in North Africa, he noted that the morale of the Italian soldiers was very low due to their defeat by the British. He made many attempts to boost their spirits during his time in North Africa.

"Italian troops had thrown away their weapons and ammunition and clambered on to overloaded vehicles in a wild attempt to get away to the west. This had led to some ugly scenes, and even to shooting. Morale was as low as it could be in all military circles in Tripoli. Most of the Italian officers had already packed their bags and were hoping for a quick return trip to Italy."

Five soldiers gather around Rommel (center) in an area protected by a wall of sandbags.

Rommel (center) converses with a German officer in a barren stretch of desert.

A convoy truck appears on the horizon (left), while a motorcycle soldier waits (right).

Rommel (right) adjusts the cuff of his sleeve while surrounded by German and Italian soldiers.

Gastone Gambara (center) was an Italian general who served briefly in North Africa. Among the Afrika Korps, he became famous for failing to appear in a decisive battle. During the battle, Rommel's commander Ludwig Crüwell asked repeatedly on the radio: "Where is Gambara?" This phrase later became a standing joke among the troops.

Gambara frequently clashed with Rommel and his staff, and eventually developed a belligerent attitude towards the Germans. He was allegedly fired after declaring openly that he wished to someday lead an Italian army against the Germans.

After being recalled to Italy, Gambara fought in Eastern Europe. He remained committed to fascist politics after the war ended. He settled briefly in Franco's Spain and later returned to Italy. He was accused of having committed war crimes while fighting in the Balkans, but was never extradited. He died in 1962.

Bastico

Rommel is surrounded by many leading Italian officers in North Africa, including Roatta (far left), Gariboldi (second from left), Teruzzi (right), and Ettore Bastico (far right).

Of all the Italian commanders he worked with, Rommel had the best opinion of Ettore Bastico.

"Marshal Bastico was a fundamentally decent man with a sober military understanding and considerable moral stamina...In general, we had worked well together and he had often supported me."

Bastico was one of the few Italian officials with no record of war crimes or crimes against civilians. After losing the battle for Libya to Allied forces, Bastico was stripped of his command and left without assignment for the rest of WWII. He spent the last years of his life studying history in Rome. He died in 1972 at the age of 96.

Rommel (center) tilts his head and smiles as he stands in a vehicle among military dignitaries at a parade.

It is likely Rommel was attending the event shown on the previous page. At the top of the rocky hillside in the background, a group of Arabs (top, left) stand on the ledge to view activities below.

Rommel saved three photos of this event in his personal collection: a picture he took and two photos of himself.

Another closeup of Rommel from the photo above.

A large crowd of soldiers gather on opposite sides of a street to await a parade.

Two Arab horsemen (left) in the village begin to ride through the dirt road towards the throng of soldiers. One Arab on a white horse (left) is dressed like a Bedouin. This image was been taken by Rommel from a high vantage point looking down at the scene below.

Rommel (center) smiles as he and other military leaders watch his troops.

Standing beside Rommel is Ugo Cavallero (right). Rommel and the Italian commanders were constantly at odds with each other. Rommel's relationship with Cavallero was particularly volatile. Cavallero had a temperamental disposition and was prone to emotional outbursts during meetings. He and Rommel's disagreements were usually very heated. Cavallero also tended to be vindictive. Following disputes with Rommel, Cavallero attempted to exact punishment; he refused to speak to Rommel, forced dialogue through intermediaries, and also used his administrative powers to remove Italian troops from Rommel's command.

After Rommel was promoted to Field Marshal, Mussolini gave Cavallero the title of "Marshal of Italy" in an attempt to prevent jealousy. Cavallero was later removed from his post. He made many enemies in the Italian government.

After Mussolini's overthrow, he sided against the Germans. Later on, he reestablished friendly relations with the Germans and was in negotiations to work on their behalf. After a meeting with German leadership, Cavallero was found shot dead in a hotel garden. It was unclear at the time whether it was a suicide or assassination. The circumstances of his death remain unsolved.

With his fingers clasped behind his back, Rommel (left) chats with a German officer in a rocky area where soldiers carry digging tools with their rifles and gear.

Rommel became very ill in North Africa. At the same time, the Afrika Korps was in crisis. Rommel wrote the following letter after he decided to remain on the battlefield despite being pressured to leave by his doctor and administrative opponents.

"It's two years today since I arrived on African soil. Two years of very hard and stubborn fighting, most of the time with a far superior enemy. On this day, I think of the gallant troops under my command, who have loyally done their duty by their country and have had faith in my leadership...I hope that my decision to remain with my troops to the end will be confirmed. You will understand my attitude. As a soldier one cannot do otherwise."

Rommel (left) leads a group of soldiers on a walk across a desert hillside.

With his goggles resting around the rim of his hat, Rommel is wearing shorts, low boots, and socks in a more comfortable attire in the desert. However, instead of wearing a short-sleeved shirt or rolling up his sleeves like some others, Rommel wears his sleeves long.

Rommel was highly critical of German military leaders who were indifferent to his troops.

"The battle in North Africa is nearing its end...The army is in no way to blame. It has fought magnificently."

"The mismanagement, the operational blunders, the prejudices, the everlasting search for scapegoats, these were now at the acute stage. And the man who paid the price was the ordinary German and Italian soldier."

"I was angry and resentful at the lack of understanding displayed by our highest command and their readiness to blame the troops at the front for their own mistakes."

A closeup of Rommel.

Rommel (left) smiles at his men as they trudge past with heavy loads on a rough dirt road.

Rommel is covered with dust. The soldiers marching along kick up dirt clouds that obscure everyone's feet and legs. The men carry a variety of gear, including a shovel, a coat, and a small metal pail along with their weapons.

When the Army's defeat became imminent, he tried to evacuate his men.

"Supplies to Africa had as good as ceased and every single man knew that the end was near—all, in fact, except our highest command. From my hospital in the Semmering (Austria), I demanded that a start should be made in bringing the troops away—of course without result. Then I asked that at least the irreplaceable people should be got out, people such as Gause, Bayerlein and Bülowius. But still no move was made. (In the event General Gause was saved by von Arnim, who sent him off to a conference in Italy; Bayerlein was sick and also flown to Italy; General Bülowius fell into enemy hands.)"

"All my efforts to save my men and get them back to the Continent had been fruitless."

Arth. Stellung by Tobruk
Sept. 41.

Rommel and his aides survey the landscape near Tobruk, Libya in September 1941.

While one Italian officer (left) gazes through binoculars and other soldiers fumble with maps, Rommel (left) has a curious expression as he looks away from them.

Rommel maintained contact with his officers after Germany lost the war in North Africa. He remained interested in their welfare after they became POWs.

"I hear that Arnim is being very decently treated by the British. So I hope that Bülowius, Seiderer, Kolbeck and all the rest of the brave boys will also find a bearable fate..."

Rommel stands with his hands behind his back (left) in a row with 10 other officers.

Without wearing his usual service cap, Rommel (center) smiles for the camera as he walks among a group of soldiers in a camp.

Several soldiers walk away from a tent in the background (center, left) and approach Rommel. The other German officer without a hat (right) wears an Iron Cross among his military decorations. He also has a gun holster within easy reach on his belt.

After he received the Victory or Death order from Berlin, Rommel and his men were in shock. He later recorded the reaction that he and his men shared.

"An overwhelming bitterness welled up in us when we saw the superlative spirit of the army, in which every man, from the highest to the lowest, knew that even the greatest effort could no longer change the course of the battle."

Soldiers (left) watch on the sideline as Rommel (center) stands among other officers.

Rommel (center, left) wears gloves and a European dress uniform in a formal military gathering as German troops arrive in Tripoli, Libya.

Lieutenant General Karl Böttcher (left) briefs Rommel (right) near several tanks.

After leaving North Africa, Rommel documented his thoughts.

"The bravery of the German and of many of the Italian troops in this battle, even in the hour of disaster, was admirable. The army had behind it a record of eighteen magnificent months...and every one of my soldiers who fought at Alamein was defending not only his homeland, but also the tradition of the Panzer Army 'Afrika.'"

"We had lost the decisive battle of the African campaign...It was even said that we had thrown away our weapons, that I was a defeatist, a pessimist in adversity and therefore largely responsible. My refusal to sit down under this constant calumny aimed at my valiant troops was to involve me later in many violent arguments and rows...The victim of it all was my army, which...fell to a man into British hands..."

Rommel and other military leaders are given a tour of outdoor facilities in North Africa.

When Germany began to lose the campaign in North Africa, Rommel's leadership came under scrutiny in Berlin. He was removed as commander, reinstated in a different capacity, and eventually forced to remain in Europe while his Afrika Korps was placed under another leader. His writings contain bitterness.

"...Army Group 'Afrika' was to be formed under my command. I received the news with mixed feelings. On the one hand, I was glad to feel that I would again be able to have some wider influence over the fate of my men...on the other hand, I was not very happy at the prospect of having to go on playing whipping-boy for the Führer's H.Q., the Commando Supremo and the Luftwaffe."

"The Army surrendered. This was followed in the Führer's H.Q. by an extraordinary collapse of morale. It was a complete surprise to them. This will be incomprehensible unless it is realized how certain people at the highest levels were waging their struggle for power on the backs of the fighting troops."

A closeup of Rommel from the original photo.

Gazing upwards with his hands behind his back, Rommel (center left) stands among 36 senior German officers in front of a chateau.

Rommel appears to be the most highly decorated officer in the group. No one else wears the Pour le Mérite, which Rommel was awarded for bravery on the battlefield during World War I. The left side of Rommel's uniform, near the pocket, also has more military decorations then the uniforms of the other officers. Other soldiers look out from the windows of the stone building.

Rommel (center) leans against tile on a rooftop where seven staff officers,
including Major Otto Heidkämper (right) have gathered in France.

Rommel returned to Europe a different man. Prior to D-Day and his assassination, he found himself opposed to the war and Germany's leaders. His writings reflect increasing hardships.

• Dec. 25, 1943 •

"I spent yesterday evening with the officers of my staff and afterwards with the men, although it's difficult to be really cheerful at the moment."

• June 10, 1944 •

"The troops of all services are fighting with the greatest doggedness and the utmost pugnacity, despite the immense material expenditure of the enemy."

• After D-Day •

"A number of generals fell in the first few days of battle, among them Falley, who was killed on the first night of the 5th-6th of June.

"My men went to their death in their thousands, without hesitation in a battle that could not be won...The sky over Germany has grown very dark."

Rommel (left) walks past rows of German soldiers and exchanges salutes with them.

In a 1944 conversation with his son, Rommel discussed D-Day and its aftermath.

"The troops behaved splendidly. During the first few days they fought among themselves for possession of rocket launchers. But then came that feeling of hopelessness... After only a few days, one of the corps commanders remained in his car when it was attacked by British low-flying aircraft. He fell back into his seat badly wounded. His A.D.C. tried to get him out before the second attack. But he held onto his seat and said: 'Leave me here. I'd rather have it that way.' The next burst of fire killed him.

"But all the courage didn't help. It was one terrible blood-letting. Sometimes we had as many casualties on one day as during the whole of the summer fighting in Africa in 1942. My nerves are pretty good, but sometimes I was near collapse. It was casualty reports, casualty reports, casualty reports, wherever you went. I have never fought with such losses. If I hadn't gone to the front nearly every day, I couldn't have stood it, having to write off literally one more regiment every day. And the worst of it is that it was all without sense or purpose. There is no longer anything we can do."

A soldier in a pith helmet (left) stands before an open vehicle and briefs Rommel (center).

Rommel (left) stands next to a rock observation post with two Italian officers and gazes into the distance.

Rommel and both Italian officers wear binoculars around their necks. A wooden chair and a bench have been placed at the base of the observation post. One Italian (right) wears a gun holster on the front of his belt.

In a speech he gave as a military instructor, Rommel summed up the qualities that made him a world-famous military commander.

"Be an example to your men, both in your duty and your private life. Never spare yourself, and let the troops see that you don't, in your endurance of fatigue and privation. Always be tactful and well-mannered and teach your subordinates to be the same. Avoid excessive sharpness or harshness of voice, which usually indicates the man who has shortcomings of his own to hide."

References

This book is not an academic work. It had its early origins as my undergraduate Honors Thesis, but evolved significantly over time. Therefore, I find it unnecessary to make a comprehensive list of all the materials that contributed to my knowledge about the topics in this book. I conducted years of in-depth research about Rommel and the war in North Africa. My writing and analyses on the subject matter are based on my own expertise. However, there are several important reference sources to which I owe credit. I cite them here as follows:

- *"The Rommel Papers,"* by Erwin Rommel (edited by B.H. Liddell-Hart and translated by Paul Findlay); with excerpts by Manfred Rommel and General Fritz Bayerlein (Da Capo Press: New York), 1953. Quoted on pp. 19, 28–29, 34–35, 42, 48, 54, 59–61, 63–65, 69, 71, 73, 79, 83–84, 88–89, 92, 100, 116, 118, 120, 122, 125, 129, 131–132, 138, 142–143, 146–147, 149, 153, 156, 160–163, 165, 167–168, 170–171, and 173.

This collection of Erwin Rommel's writings and personal letters was published shortly after World War II. At the request of the Rommel family, it was edited by British military historian B.H. Liddell-Hart, whose work Rommel had admired during his lifetime. Various eyewitnesses contributed to the book, including Manfred Rommel, his mother Lucie, and General Fritz Bayerlein, who had fought alongside Rommel in North Africa.

- *"Rommel, the Desert Fox,"* by Desmond Young (Harper & Brothers: New York), 1950.

Desmond Young was a British officer in WWII who was captured as a POW by German troops in North Africa. He experienced the desert war firsthand and met Erwin Rommel in person. While gathering material for this book, he personally interviewed Rommel's family members, friends, fellow soldiers, and other eyewitnesses; gained access to original documents; and recorded firsthand testimony.

- *"The Ciano Diaries,"* by Count Galeazzo Ciano (Doubleday: Garden City, N.Y.), 1946.

Count Galeazzo Ciano was Mussolini's son-in-law and Minister of Foreign Affairs in fascist Italy. His position placed him in close proximity to many military and political figures in both Italy and Germany. Ciano kept diaries, in which he recounted events he witnessed. His writings contained scathing and sarcastic commentary. Ciano was eventually accused of treason by Mussolini and executed by a firing squad. Ciano's writings provide insight into the inner-workings of Italy's military and government during WWII.

- *"Rommel's Army in Africa,"* by Dal McGuirk (Motorbooks International Publishers: Osceola, Wisc.), 1993.

This book provides a detailed view of what life was like for Rommel's men in the desert. Dal McGuirk collected letters and artifacts from the war in North Africa and corresponded with Afrika Korps veterans, who provided him with insights.

ENGLISH TRANSLATIONS OF ROMMEL'S WRITINGS

The excerpts quoted from Rommel's writings included in this book come from *"The Rommel Papers."* Because Findlay's English translations of Rommel's writings in German were made more than 50 years ago, I occasionally substituted some antiquated British words with modern English equivalents.

About the Author: Zita Steele

Photo by Noël Fletcher

Zita Steele is a novelist and artist from the Southwestern state of New Mexico. She writes both fiction and nonfiction. She has expertise in criminology, cybercrime, and international relations. She loves foreign languages and cultures. Zita's stories often involve history and international themes. She likes to create characters who must cross cultural divides to understand each other.

She is currently working on the third volume of the nonfiction *"Erwin Rommel: Photographer"* series and also on fiction projects.

Other Books by Zita Steele

Erwin Rommel Photographer–Vol. 1 A Survey

by ERWIN ROMMEL and ZITA STEELE

Take a journey behind the camera of a world-famous military commander. Experience WWII firsthand from Field Marshal Rommel's private photo collection, seized by U.S. forces in 1945. View 340+ images, including photos Rommel took during campaigns in France and North Africa and others he collected. Included are Rommel's personal photos of family and friends. The photos are digitally restored for detail. Some are accompanied by Rommel's own handwritten photo captions. Author/artist Zita Steele uses her knowledge of German language and culture, with in-depth research about Rommel and his campaigns, to provide context for the photos. Zita also analyzes patterns in Rommel's photography to shed a light on the artistic personality of this notable military leader.

Edge of Suspicion *by* ZITA STEELE

Justin Moon of South Korea is the world's top private eye. He travels to Singapore to catch an elusive cybercriminal. The pay is lucrative. His client is an attractive blonde CEO. It should be the easiest job in his career. Things get complicated with the arrival of Okada, a mysterious drifter with a mission of revenge. As Moon tries to solve the mystery, he uncovers a tangled maze of deceit. Each new clue leads him in an unpredictable direction. A deadly game of cat-and-mouse begins. Featuring over 100 photos, *Edge of Suspicion* is both an exciting story and a work of art.

Envoy: Rule of Silence *by* ZITA STEELE

Take a journey into a thrilling world of secrets and lies in modern-day Europe. Polish ex-secret policeman Michal Krynski is tired of working as a double agent for France's security bureau. His last mission: to track down a runaway DJ. As he travels to the strange island of Malta, Krynski plots revenge against the system that ruined his life. Will he catch the DJ or kill him? Zita Steele is a novelist and artist. She writes with an expertise in criminology, cybercrime, and international relations. She creates her own illustrations.

More Books from Fletcher & Co. Publishers

**FLETCHER & CO.
PUBLISHERS**

Every book is a journey

Every book is a journey. Fletcher & Co. Publishers is an independent, art-house publishing company. We use new media and graphic design techniques to transport you into the world of the novel.

Our books aren't just written words. They're experiences: international cultures, art, suspense, history, and adventure.

Watch our video trailers on our YouTube and Vimeo channels to preview each book, see interesting images, and learn more about our newest releases. Visit us on our website or Facebook to find out about our latest news.

Coming Soon!

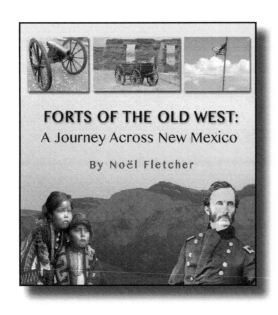

**Forts of the Old West:
A Journey across
New Mexico**

by Noël Fletcher

Take a journey into the forgotten forts of the Wild West and discover their dark secrets.

The Strange Side of War *by* SARAH MACNAUGHTAN & NOËL FLETCHER

THE STRANGE SIDE
of WAR

A Woman's WWI Diary

By Sarah Macnaughtan
Introduced & Edited by Noël Fletcher

Take a journey across the dangerous battlefields of a world at war. Accompany Scottish novelist Sarah Macnaughtan as she volunteers alongside British humanitarian groups to alleviate the suffering in war-torn lands. Her many adventures tell unique stories of tragedy and triumph, taking readers on an unforgettable journey from the trenches of Belgium to the distant frontiers of Persia and tsarist Russia.

Author/editor Noël Fletcher provides new historical context that brings Sarah's story to life and helps readers to remember the bravery and sacrifice of those who died. Illustrated with 130+ rare photos and propaganda posters from World War I, this important work features historical insights about the people and places involved in the conflict.

New Mexico Ghosts and Haunting Images *by* ARIELA DESOLINA

Let explorer-photographer Ariela Desolina spirit you away to New Mexico, where haunting ruins - some with ghostly inhabitants - will capture your imagination. With photos from the St. James Hotel, a notorious hangout of Western outlaws and gamblers, and other mysterious locations.

Mysterious shapes and ghostly forms (undetected when the pictures were taken) sometimes appear in her photos. This collection features photos of the notorious St. James Hotel in Cimarron, a famous haunt of outlaws and gamblers and the haunted ruins of the Kelly Mine, once among the richest old gold & silver mines in the Southwest.

NEW MEXICO GHOSTS &
HAUNTING IMAGES

by Ariela Desolina

River of My Ancestors: The Rio Grande in Pictures *by* NOËL FLETCHER

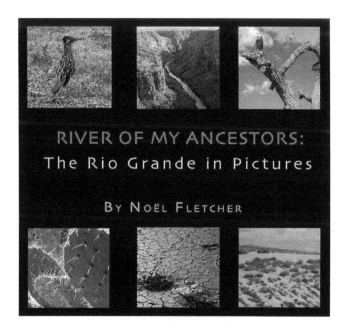

Take a journey along the wild and rugged Rio Grande. Beautiful pictures capture the essence of the famous river and its importance in the arid Southwest. Native New Mexican author and photographer Noël Fletcher provides family stories and insights about frontier life.

Follow the Rio Grande through deserts, wetlands, and rocky cliffs. Experience natural wonders, including volcanic lands and river rapids, and encounter wildlife such as snakes, wolves, cranes, and bighorn sheep.

With 180+ striking color photos, the book features:

- Author biography
- Wild West history
- Interesting facts about New Mexico, local culture, and life along the Rio Grande
- the world's largest cottonwood forest
- the legendary Rio Grande Gorge
- Spanish colonial irrigation systems
- Bosque del Apache National Wildlife Refuge
- Unique wildlife and plants
- Oral tradition from Spanish settlers and family stories

This captivating book combines vivid photos and the written word to tell a living history of the famous Rio Grande and the beautiful desert land of New Mexico.

Two Years in the Forbidden City *by* PRINCESS DER LING

This true story was the first eyewitness account of the Imperial Court written by a Chinese aristocrat for Western readers. It provides an up-close view of the notorious Dowager Empress Tzu-hsi in her final years. Enhanced with rich imagery and additional historical notes, it includes interesting historical details and photos about China's infamous Dowager Empress, the Boxer Rebellion and the Imperial Court.

It is illustrated with 100+ historical photographs, illustrations, and paintings from the late 1800s to early 1900s. Author/editor Noël Fletcher that provides context for this book in modern Chinese history.

Lantern of the Wicked *by* CHARLES CLEMENT

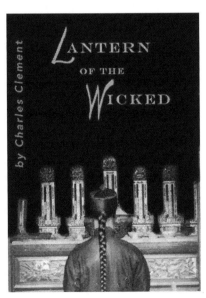

In the decadent and dangerous Shanghai of 1929, someone is spying for the Japanese, and the International Settlement's British police are on the hunt.

Now, in the midst of the Mid-Autumn Moon Festival, American aviator Jack "Ace" Jordan becomes the prime suspect.

A thrilling narrative blending fact, fiction and rare photographs, *Lantern of the Wicked* creates an atmospheric window into the complexity and dark grandeur of the colonial Orient in this gripping historical mystery.

It's All Good: A Story of Love, Loss & Hope

by Neil Candelaria

Take a journey of self-discovery and healing. Learn about the short but unforgettable life of Casey Jang-Joon Candelaria, a boy adopted from Busan, South Korea, who found a loving family and home in the American Southwest. His death at age 25 left his family shocked and devastated. Author Neil Candelaria, Casey's father, shares his experiences as a parent coping with the loss of a child. A former criminal court judge, Neil relates how he overcame his reliance on physical proof and came to believe in life after death.

The Spy *by* JAMES FENIMORE COOPER

During the dark days of the Revolutionary War, America struggles for nationhood. Meanwhile, in the shadows, a spy is trading secrets of vital importance to the cause – but for whose side? Colonials and loyalists play a game of cloak and daggers in a classic historical of action and adventure.

Our edition features 30+ color photographs, chapter titles, and illuminating notations, designed to give you a front-seat experience.

This was the first major fiction novel on espionage ever written and published in America.

Mystery of the Yellow Room *by* GASTON LEROUX

News of a strange crime spreads like wildfire in Paris. Someone has attempted to murder the daughter of a brilliant scientist. But nobody can explain how the murderer got in and out of the locked room of her isolated country home. Only Joseph Rouletabille, an impatient young journalist, has the genius to solve this crime.

Written by the author of *"The Phantom of the Opera,"* this novel was published in 1907 as France's reply to Sherlock Holmes. Our edition has adapted text from archaic Victorian to standard English. It also features updated maps and is illustrated with 30+ historical paintings and illustrations from 19th century France.

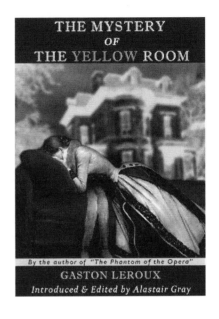

Made in the USA
Middletown, DE
04 August 2018